Understanding
Boat Wiring

OTHER BOOKS BY JOHN C. PAYNE

The Fisherman's Electrical Manual

The Great Cruising Cookbook: An International Galley Guide

Marine Electrical and Electronics Bible, Second Edition

Motorboat Electrical and Electronics Manual

Understanding Boat Batteries and Battery Charging

Understanding
Boat Wiring

JOHN C. PAYNE

SHERIDAN HOUSE

This edition first published 2003 by
Sheridan House Inc.
145 Palisade Street,
Dobbs Ferry, NY 10522

Library of Congress Cataloging-In-Publication Data
Payne, John C.
        Understanding boat wiring /John C. Payne.
            p. cm.
    ISBN 1-57409-163-8 (alk. paper)
        1. Boats and boating-Electric equipment. 2. Electric
    wiring. I. Title.
    VM325. P3924 2003
    623.8'503-dc21   2002154370

Printed in the United States of America
ISBN 1-57409-163-8

Contents

1. BOAT WIRING STANDARDS

Approximately 85% of vessel failures can be attributed to improper connections, terminations or incorrectly installed cables. Using accepted wiring practices can eliminate failures. Unfortunately the common attitude is to treat vessel low voltage systems like automotive installations, and the high failure rate on boats reflects this attitude. Exposure of DC electrical systems to water may cause fire, shock and catastrophic damage.

Circuit Arrangement

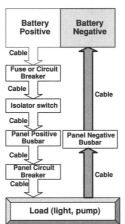

What electrical standards are required?

Electrical systems should be installed to comply with one of the principal standards or recommendations in use, and most standards are similar. Rules are relatively expensive to purchase and many people are also intimidated by the complexities. Most rules are difficult to interpret, as they are not written in plain language, but consist of technical jargon and assume considerable electrical knowledge. The following chapters cover many of the requirements contained in the various recommendations. They represent best practice and will help in getting your boat installation to a similar level. Where you are required to use standards, a copy of the relevant standard is essential.

- **American Boat and Yacht Council (ABYC).** *Standards and Recommended Practices for Small Craft. E-9 DC Electrical Systems.*
  These are voluntary standards and recommendations that are widely used by many US boat builders.

- **The United States Coast Guard.** *Title 33, CFR 183 Subpart I, Section 183.*
  These contain mandatory requirements for electrical systems on boats.

- **NFPA 302,** *Fire Protection Standard for Pleasure and Commercial Motor Craft,* 1994 Edition.
  This standard is approved by the American National Standards Institute and is applicable to motorboat installations.

- **Australia.** *Uniform Shipping Law (USL) Code. Section 9.*
  These rules are for commercial boats being built in survey.

- **European Recreational Craft Directive (UK and Europe).**
  These RCD standards are now virtually mandatory on new construction boats in European Union countries and include the following: EN ISO 10133. *Small Craft -Electrical systems-Extra-low-voltage DC installations,* 1994; EN 28846:1993 *Electrical devices-Protection against ignition of surrounding flammable gas;* EN28849:1993 *Electrically operated bilge pumps;* EN ISO 9097:1994 *Electric fans;* prEN ISO 16180 *Inboard diesel engines Engine mounted fuel and electrical components.*

- **Lloyd's Register.** *Rules and Regulations for the Classification of Yachts and Small Craft.* These rules are mainly used for boats being built in survey.

Large Panel

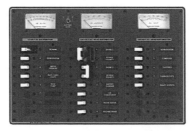

2. BASIC ELECTRICAL PRINCIPLES

Introduction

The Greeks discovered electricity around 500 BC, when static charges were raised on the surface of amber. The word electric comes from the Greek meaning "to be like amber." There have been many later discoveries as various scientists developed these first principles. Some basic electrical principles need to be understood if a system is to be designed, installed, maintained and repaired. The majority of problems can be usually traced to these basic rules.

What is Ohm's law?

This is a fundamental electrical law that was formulated by Georg Simon Ohm, a nineteenth-century German physicist. One volt is the ElectroMotive Force (EMF) required to move a current of one amp through a resistance of one ohm. Ohm's law is expressed in some very simple formulae which are the basis of all simple electrical circuits:

Voltage V (Volts) = Current (Amps) x Resistance (Ohms)
Current I (Amps) = Voltage (Volts) divided by Resistance
Resistance (Ohms) = Voltage divided by Current
Power (Watts) = Voltage multiplied by Current
Current (Amps) = Power (Watts) divided by Voltage

Example 1. A circuit has a voltage of 12 volts, a current of 5 amps. What is the resistance? R = 12÷5, R = 2.4 ohms.

Example 2. A circuit has a current of 10 amps, a resistance of 1.5 ohms. What is the voltage? V = 10÷1.5, V = 6.7 volts.

Using the Ohm's law triangle

This is an easy method for working out the relationship between each part and making calculations. Place a finger over the unknown quantity in the triangle to determine what calculation is required.

Example 1: It is known that the resistance is 20 ohms, and that a voltage of 14.0 volts is across the resistance. The current passing through the resistor is unknown. Place a finger over I for current in the triangle. This leaves the voltage V over the Resistance R.
Current A therefore equals V÷R = 14÷20 = 0.7 amps

Example 2: It is known there is a voltage of 12.5 volts and a current of 1.7 amps is flowing. The resistance of the circuit or device is unknown. Place a finger over the R for resistance in the triangle. This leaves the voltage V over the Current I.
Resistance R therefore equals V÷I = 12.5÷1.7 = 7.35 ohms

Ohm's Law Triangle

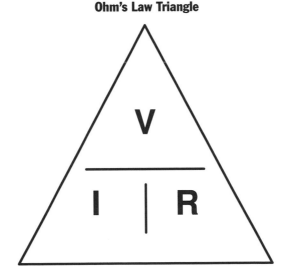

What is a conductor?

Conductors are made from materials that allow the easy flow of electrons, or conduct electricity easily. These include metals such as silver, copper, gold and aluminum which have the lowest resistances and the highest conductivity. Every conductor has a different resistance value, varying usually with the temperature. In electrical systems copper is used for most purposes.

What is insulation?

Insulators are made from materials that will resist or prevent the flow of electricity, and have a high resistance. Typical insulators are glass, wood, PVC and rubber. Insulating materials are used to cover conductors and terminal blocks. Insulating materials are also used to support conductors and terminals.

Conductor and Insulation

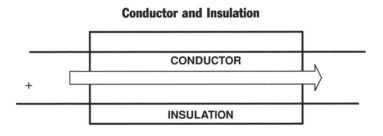

What is resistance?

This is defined as the opposition to the flow of current in an electric circuit. It is defined by Ohm's law. The unit of resistance is the ohm, and the symbol of the Greek letter Omega (Ω). The basic equation for resistance equals the voltage divided by the current, i.e. $V \div I$. The ratio of a voltage through a conductor to a current flowing in it is constant, and is equal to the resistance of the conductor. When current flows through a resistor, heat is generated and is often expressed as dissipated power in watts. This is the basis of all heating elements. Resistance is also directly proportional to the length (L) of a conductor, with resistance varying inversely with the cross-sectional area (CSA).

What is voltage?

Electromotive Force (EMF) is measured in volts. EMF can be generated by static charges, by chemical reaction such as in an electromechanical cell or battery, or by moving a conductor in a magnetic field. The volt is the unit of electrical pressure, and is the force required to cause a current to flow against a resistance. The basic equation is $E = I \times R$. This is defined as a unit of electric potential or electromotive force. It is the difference of electric potential that exists between two points on a conducting wire carrying a constant current of one ampere when the power dissipated between the points is one watt. Voltage at a higher potential flows to one of lower potential, that is positive to negative.

What is current?

This is defined as the rate of electric current flow. This is the movement of electrons through a conductor in a single direction only. In general when the voltage increases in the circuit the current also increases. The unit of current is the ampere or Amp.

What is power?

The unit of power which for a DC circuit is equal to the voltage multiplied by the current, i.e. P (watts) = V x I. The standard unit is the watt, symbolized by W. In a DC circuit, a source of V volts, delivering I amperes, produces P watts. When a current of x amperes passes through a resistance of x ohms, then the power in watts dissipated or converted by that component is given by P = I^2R. To work out current from the power or watt rating use the formula of I = P÷V. Example: Power rating is 135 watts, the voltage is 24 volts. I = 135÷24 = 5.6 amps.

What is an electrical circuit?

A circuit consists of all the various elements which are a source of electrical power, the wire and cable that carry the electrical power, the load or item that is being powered such as a light, the various isolation and protection devices such as switches, and fuses.

What is a series circuit?

This is a circuit which has devices connected end to end, that is positive of one to the negative of another. These circuit resistances are simply added together. In boat systems, the only common series circuit is connecting batteries in series to create a higher voltage. In practice the load has a resistance, and also the cables and protective devices so that in effect there are two resistances. Resistances connected in series carry the same current.

What is a parallel circuit?

This is a circuit where each load or device is connected in parallel. In most boat systems, the most common parallel connection is the lights. Each light is connected across the same positive and negative electrical supply. Resistances connected in parallel carry the same voltage. If two resistances of the same value are connected in parallel the resistance is halved.

How to use a multimeter

The majority of tests can be carried out using a multimeter. A multimeter, as the name suggests, is able to perform a range of electrical measurements. There are two types of multimeters, analog and digital. An analog meter has a needle to show the readings. The digital meter (DMM) displays the test values numerically on a display. Manual ranging meters require selection of measurement ranges, and auto-ranging types select the best measurement range.

Multimeter
Courtesy Technika

How to measure voltage

It is the most useful of all measurements, either to detect that it is present or to precisely measure voltage levels. I perform 95% of all my troubleshooting on complex oil rigs and commercial vessels with this function alone. The voltmeter is connected across the supply or equipment, which is negative probe to negative and positive-to-positive to measure the voltage potential between the two. Reversal of probes will simply show a negative reading. If the DMM is not auto-ranging, set the scale to the one that exceeds the expected or operating voltage of the circuit under test. To analyze results:

- If the voltage is missing, this indicates that the circuit supply is switched off, or the circuit is possibly broken, such as a connection or a wire (positive or negative), or a faulty switch or circuit breaker.

- If the voltage is low, this indicates that the supply voltage to circuit from the battery is low, or that additional resistance is in the circuit, such as faulty connection.

How to make continuity tests

The continuity test requires the use of the Resistance ohm setting. It is simply to test whether a circuit is closed or open. Many multimeters also incorporate a beeper to indicate a closed condition. Power must be switched off before testing. Set the scale to one of the megohm ranges. Touch the probes together to verify operation, and then place the probes on each wire of the circuit under test. What you are looking for is a simple over-range reading if the circuit is open, and low or no resistance if it is closed.

How to measure resistance

If DMM is not auto-ranging, set the range switch to the circuit under test, typically the 20 Ohm range is used. Turn off circuit power, and discharge any capacitors. When testing, do not touch probes with fingers as this may alter readings. Prior to testing, touch the probes together to see that the meter reads zero.

How to make current tests

The ammeter function of a multimeter is rarely used or required. The switchboard ammeter normally can be used for all measurements. The ammeter is always connected in series with a circuit, as it is a measurement of current passing through the cable. The circuit should be switched off before inserting the ammeter in circuit. Most DMM have maximum DC measurement ratings of 10 amps only. It is a little used function.

How to maintain your meter

Look after your meter. Do not drop the meter or get it wet. There are a few basics that ensure reliability and safety. Ensure that probes are in good condition. On many probes the tips sometimes rotate out, and a probe may came out and short across the terminals under test. Another problem is the solder connections of test leads breaking away due to twisting and movement. Cables should be kept clean and insulation undamaged. Cables can age and crack. If a cable is damaged replace it. Do not attempt to test higher voltages, in particular AC voltages, if the cables are damaged. People have received severe shocks or been killed because of faulty leads. Replace the internal battery every 12 months, or at least carry a spare. Many meters will have a low battery warning function.

3. ABOUT SYSTEM VOLTAGES

Which voltage should be used?

It is quite common to see boats having both 12- and 24-volt systems in use, some older boats have 32 volts and the newer 36- (42-) volt systems will also have the same factors. Effectively where two voltages are used they should be treated as two entirely separate systems. In polarized ground systems the negatives will be at the same potential. This will mean two alternators and two battery banks. The merits of 24 volts for heavy current consumption equipment such as inverters and windlasses are obvious, because the cables are half the size and weight of 12-volt systems. In many cases electronics will be able to operate on 24 volts without modification.

About 12 (14) volt systems

The 12-volt system is the most common system. Because of automotive influence, which have led to a large range of equipment being available, 12-volt systems are used to power most boat electrical and electronics equipment. It is also possible to purchase virtually any appliance rated for 12 volts. The charge voltage is 14 volts.

About 24 (28) volt systems

This system is common, especially in commercial applications. It has the advantage of lower physical equipment sizes, cabling, and control gear and in addition voltage drops are not as critical. The charge voltage is 28 volts. Because much equipment is commonly 12 volts, a DC-DC converter must be used to step down to 12 volts. Although complicating the system, this isolates sensitive electronics equipment from the surge and spike-prone power system.

About 32 volt systems

This is an old system voltage that is still found on some boats. 8-volt batteries are still manufactured and four batteries are connected in series to make 32 volts.

About 36 (42) volt systems

This is a new automotive voltage to be introduced in 2002 and which may or may not eventually transfer to boats. It consists of three 12-volt batteries connected in series to make up 36 volts, and the charge voltage is 42 volts.

About 48 volt systems

This system voltage is now starting to become more prevalent, in particular for powering thrusters. Where this voltage is used, two 24-volt battery banks are connected in series using a switch and relay unit to make 48 volts.

How are voltages converted?

In many boats, a mix of voltages requires the use of DC converters to step down from 24 to 12 volts, and, if implemented on boats, the same will be required for 42-volt systems. Good quality converters have a stabilized output of around 13.6 (27.2) volts. Typical power consumption of a converter without a load connected is approximately 40 to 50 milliamps, so there will always be a battery drain. The converter should ideally have an isolation switch on the input side. Most converters are installed with automatic thermal shutdown, short circuit fuse protection, current limiting and reverse polarity protection. Good ventilation is essential and converters should be mounted vertically so that fins are also vertical to facilitate convection. Sufficient clearance must be allowed between top and bottom.

4. HOW TO PLAN BOAT WIRING

How is a boat wired or rewired

The average boat now has many systems and equipment installed. Planning the installation requires a carefully considered systems approach. In the majority of cases, systems are over complicated and follow no accepted electrical practice. They have inherent system problems that are only overcome with costly total rewires. Use the KISS principle, do it once, and do it right!

How to make a wiring plan

Each boat requires a complete wiring diagram showing all the wiring, equipment and systems installed. The diagram should include Equipment Identification, Equipment Current Rating, Cable Sizes, Circuit Breaker and Fuse Ratings, and Circuit Identification.

1. Make a plan of your boat, including all bulkheads and spaces and then locating every item of equipment on it. Write down the equipment identification name.

2. Write down the current draw in amps for each item of equipment. Enter these into the battery calculation tables. This will allow calculations to be made on the required battery capacity and charging requirements.

3. Draw in the proposed cable route for each item of equipment, showing bulkheads, decks or other obstructions. Where the cable will be routed within bilge areas, or exposed to mechanical damage, use an alternative route.

4. Determine the cable size by using the current draw and calculating the voltage drop within the circuit.

5. Determine the required protection required, and then enter the circuit breaker or fuse rating for the circuit and assign a circuit number. Use a logical sequence, such as switchboard left hand vertical row is No. 1 downward and so on.

Boat Outline - Planning

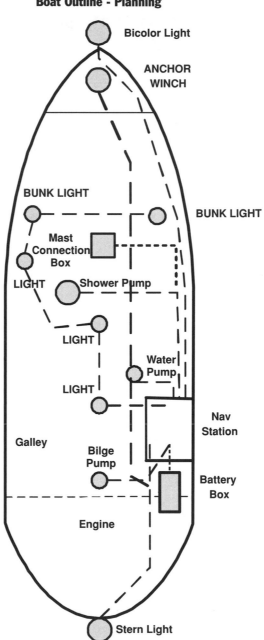

How does hull material affect wiring?

Hull material has important implications with respect to the wiring configuration. A steel or aluminum boat hull is conductive and will require an insulated return electrical system, with careful consideration of corrosion, grounding and lightning protection. On a new boat, engines with isolated return electrical systems can be purchased. A fiberglass or timber boat may have a simpler wiring system, corrosion being only a limited factor.

How important is the boat size?

The boat size affects the length of cable runs, with greater cable sizes and weights and increased voltage drop problems. It is an important factor on canal boats and barges, and on catamarans. This affects voltages, and large boats have a real case for selecting 24 volts. The voltage drop problems are reduced, the battery weight and sizes for a given capacity are less, and the weight and size of equipment is generally reduced. Larger powerboats also have a greater level of accommodation, and therefore more people are often aboard, (the parties are longer!) putting greater demands on batteries through increased lighting, larger electric refrigeration, with consequential increased requirements on charging. This lifestyle factor, largely ignored by boat builders, is important.

Installing equipment in hazardous areas

Electrical equipment and cables should not be installed within any compartment or space that may contain equipment or systems liable to emit explosive gases and vapors. This may include spark ignition engine fuel systems, LPG installations or flooded cell battery installations. Where any equipment or fittings are to be installed, they must be ignition protected in accordance with the appropriate national standards.

Locating electronics equipment correctly

Where practicable, navigation electronics equipment, control modules, processors and related components should be located clear of cable looms and aerial cables to prevent interference. This should include all radar, satellite communications and television systems, cellular telephones, ham, VHF, SSB/HF tuners and control units and aerial lead-in wires. Autopilots are prone to interference causing major uncontrolled course alterations: in fact an autopilot severely affected by a cell phone led to a fatal accident. A minimum distance of 36" (1m) is recommended.

Equipment Clearances

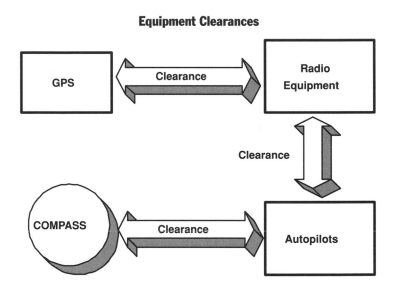

What is a two-wire insulated configuration?

The two-wire insulated return system is preferred for all steel and alloy boats. This configuration has no part of the circuit, in particular the negative, connected to any ground or equipment. The system is totally isolated or floating, and this includes all circuit negative conductors, engine sensors, starter motors and alternators. The hull must never be used as a negative return path. In the two-wire insulated system, each outgoing circuit positive and negative supply circuit has a double pole short circuit protection and an isolation device installed. This may be incorporated within a single trip free circuit breaker. The isolator has to be rated for the maximum current of the circuit. In this configuration, a short circuit between positive and ground will not cause a short circuit or systems failure. A short circuit between negative and ground will have no effect. A short between positive and negative will cause maximum short circuit current to flow.

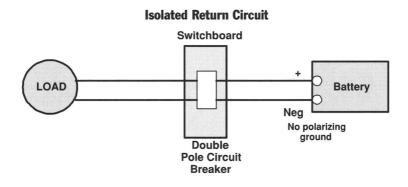

Isolated Return Circuit

What is two-wire one-pole grounded wiring configuration?

The two-wire with one pole grounded system is preferred for fiberglass and timber vessels. This is also called a polarized system. It is the most common configuration, and holds the negative at ground potential by connection of the battery negative to ground, or in most cases the mass of the engine block. In most installations, the main negative to the engine polarizes the system, as the engine mass and connected parts such as shaft provide the ground plane. There should only be one polarizing ground conductor. In a two-wire, one-pole grounded systems, each outgoing circuit positive supply circuit requires a short circuit protection and isolation device installed. This may be incorporated within a single trip free circuit breaker. The earthed pole should not have any protective device installed. In this configuration, a short circuit between positive and ground will cause maximum short circuit current. A short circuit between negative and ground will have no effect. A short between positive and negative will cause maximum short circuit current to flow. The single pole circuit breaker will break positive polarity only.

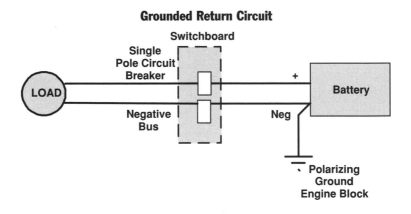

Grounded Return Circuit

What is a distributed wiring system?

In this arrangement the power supply is broken down into a system of sub-panels, and it is becoming increasingly preferable on larger boats. There are many advantages over a centralized system, including the separation of potentially interactive equipment such as pumps and electronics. Other options include intelligent control systems that have remote control of circuits, and systems such as touch screens to switch circuits. Separation enables a reduction in the number of cables radiating throughout the vessel from the main panel to areas of equipment concentration. This is a cause of RFI and considerably greater quantity of cable. Most distributed systems run all the sub-circuits from the central panel, with each circuit having a circuit breaker to protect it. In the illustrated example only essential services are kept along with metering on the main panel. Lighting panels can be located anywhere practicable; once the circuits are on, lights are switched locally.

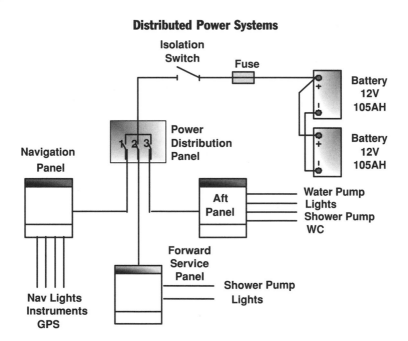

Distributed Power Systems

What about the new 42-volt systems?

These will require a two-voltage system in a distributed configuration. The 42-volt system will be supplied by a high output alternator rated at up to 4kW, which is around 100 amps. The output will supply a 36-volt battery bank, and a high current consumer panel. Typically this panel will supply the thrusters, hydraulic power unit (HPU), anchor windlass, winches, and DC electric propulsion. As these consumers are normally powered up when an engine is running, the full 42 volts are available. A power management system will be required. Low power consumers will be powered from a separate 12- or ideally a 24-volt battery system. These consumers will comprise lighting loads, electronics, and auxiliary power circuits.

42 Volt Electrical Systems

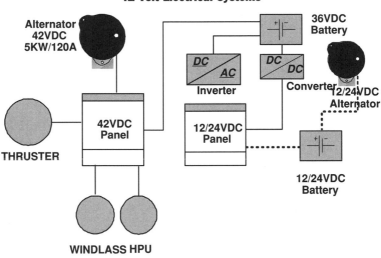

How are wire sizes calculated?

Conductors must be selected based on the maximum current demand or ampacity of the circuit. All cables have a nominal cross-sectional area (CSA) and current carrying capacity. Standards specify nominal capacities for a range of cross-sectional areas and temperature ranges. The rule of thumb value is 6-8 amps per mm^2.

Typical DC Cable Nominal Ratings and Data

AWG	CM Circ mil	CSA	Area mm^2	DC amps
20	1024	.52	.5	1.5-2
18	1624	.82	.75	2-3
17	2052	1.04	1	3-4
16	2581	1.31		
15	3260	1.65	1.5	4.5-6
14	4109	2.08		
13	5184	2.63	2.5	7.5-10
12	6529	3.31		
11	8226	4.17	4	12-16
10	10384	5.26		
9	13087	6.63	6	18-24
8	16512	8.37		
7	20822	10.55	10	30-40
6	26244	13.30		
5	33088	16.77	16	48-64
4	41738	21.15		
3	52624	26.66	25	75-100
2	66358	33.62	35	105-140
1	83694	42.41		
0	105625	53.52	50	150-200
2/0	133079	67.43	70	210-280
3/0	167772	85.01		
4/0	211600	107.21	95	285-380

How to calculate cable and wire sizes

In the US the more common method of calculating cable current ratings, voltage drop and ampacity is the use of charts and tables. The charts are used either for 3% voltage drop for navigation lights, main power feeds and electronics, and 10% for other circuits. The vertical scale has the current in amps and the horizontal scale is the total length of the circuit. In DC circuits both the positive circuit wire to the equipment and the negative must be added to get the total length. The following formula can be used.

$$CM = K \times I \times L/E$$

CM = Circular Mil area of the conductors
K = 10.75 (a copper resistance constant per mil-foot)
I = Current in amps
L = Conductor length in feet
E = Voltage drop at the load in volts

Example 1: To find CM cable size for 5% voltage drop in a navigation light circuit with 2 amp lamp, conductor distance is 100 feet to mast head.

$$CM = 10.75 \times 2 \times \frac{200}{.6}$$

CM = 7167 from table closest is 11 AWG

Example 2: To find voltage drop in a navigation light circuit with 2 amp lamp, conductor supply length is 100 feet to mast head using 11 AWG wire.

$$E = K \times I \times L/CM$$
$$E = 0.75 \times 2 \times 200/ 8226$$
Volt drop = .52 volts

How to calculate voltage drop

Voltage drop is always a consideration when installing electrical circuits. Unfortunately, many voltage drop problems are created by the poor practice of trying to install the smallest cables and wiring sizes possible. The ABYC sets a recommendation of 3% and 10% for circuits. I recommend having a maximum acceptable voltage drop in 12-volt systems of 5% or 0.6 volt for all circuits, and aim for 3% on the ABYC required equipment such as navigation lights and electronics. A 10% limit is really excessive for most equipment, in particular circuits such as bilge pumps. The voltage drop problem is prevalent in starting and charging systems, thrusters, windlasses and trolling motors. The formula is specified in ISO Standard 10133, Annex A.2.

$$\text{Voltage Drop at Load (volts)} = \frac{0.0164 \times I \times L}{S}$$

I = is load current in amperes
L = is cable length in meters, positive to load and back to negative.
S = is conductor cross-sectional area, in square millimeters

Example: For an anchor windlass, these cable sizes must be calculated at normal working and peak loads. As the calculations show, a larger cable size ensures less voltage drop and fewer line losses. Working-load current = 85 amps, cable run = 12 meters, CSA = 35 mm² rating = 125 amps.

$$\text{Drop at 85 amps} = \frac{0.0164 \times 85 \times 24}{35}$$
$$= 0.96 \text{ V } (35 \text{ mm}^2) \; 0.67 \text{ V } (50 \text{ mm}^2)$$

$$\text{Drop at 125 amps} = \frac{0.0164 \times 125 \times 24}{35}$$
$$= 1.41\text{V } (35 \text{ mm}^2) \; 0.98 \text{ V } (50 \text{ mm}^2)$$

Selecting cable types

Double insulated cables should be used on all circuits to ensure insulation integrity. Insulation is temperature rated, which has important implications with respect to cable ratings. In most boats PVC insulated and PVC sheathed cables rated at 170°F (75°C) are used. For classification societies, ship wiring cables that use Butyl Rubber, CSP, EPR or other insulating materials are specified. They have higher temperature ratings and higher current carrying capacities.

Compensating for high temperatures

Ambient temperatures exceeding the rated temperature of the cable should be de-rated by a factor of 0.05 for each 9°F (5°C) above. Temperature reference is typically 75°F (25°C), although 85°F (30°C) is used. All cable current carrying capacities are subject to de-rating factors. In any installation where the temperature exceeds the nominal value, the continuous current carrying capacity of the cable is reduced. This is important in engine spaces. Where the temperature exceeds 120°F (50°C) the de-rated capacity is the nominal capacity multiplied by 0.75, and some recommendations de-rate by 15%. Consult actual cable manufacturers' ratings for accuracy.

What are minimum conductor sizes?

The minimum conductor size to be used should be 16 AWG (1.0 mm^2), and this is also an ABYC requirement. All cables in any circuit must always be the same rating. I would recommend that all conductor sizes be standardized to 12-14 AWG (2.5 mm^2) for all general circuits subject to considering the current and volt drop requirements. Cable is cheaper to purchase by the roll.

How are wires coded or identified?

In most parts of the world, conductors are identified as red for the positive conductor, and black for the negative conductor. Numbered cores are also an acceptable alternative; the numeral 1 should be used as the positive and numeral 2 used for negative. Most places in the world modified AC systems coding and moved to IEC standards to avoid confusion, leaving black and red for DC, and brown and light blue for AC. This did not happen in the US, and was further complicated with the ABYC nominating yellow as a negative polarity color. In many countries yellow is also a primary AC phase, or control or switching wire in AC systems, so be careful. The best way to avoid any confusion is to ensure that AC and DC are not installed within close proximity. ABYC defines a color code system which some boat makers follow.

ABYC Wiring Color Code Recommendations

Color	Circuit Function
Yellow/red	Starting circuits
Brown/yellow	Bilge blowers
Dark grey	Navigation lights/ tachometer
Orange	Accessory feeds
Brown	Pumps
Purple	Instruments
Dark blue	Oil pressure
Tan	Water temperature
Pink	Fuel gauge sender

What type of conductor is best?

Conductors should be stranded and tinned copper. Stranding gives flexibility. Type 2 and Type 3 are preferred. Single solid conductor wire such as that used for house wiring should not be used. When untinned copper is exposed to saltwater spray or moisture, it will very quickly corrode, degrade and fail. The argument speaking against the installation of tinned copper is cost. The price differential is typically 30% greater and the reliability (and boat resale increase) advantages far outweigh the lower priced plain copper conductor. In some vessels, conduits are installed with single insulated wires; however they are easily nicked or damaged during installation and it is highly recommended that double-insulated cables be used. In some parts of the installation such as bilge pump circuits a triplex wire can be used and 5-conductor multi-cables for mast lighting.

What are shielded conductors?

These cables generally have the two conductors twisted, and are called twisted pairs. They have a woven copper wire shield surrounding them to prevent radiating electromagnetic interference. The shield is then grounded at one end.

What are duty cycle ratings?

Where cables carry large currents for short time durations, they should be used subject to duty cycles. Heavy current carrying cables such as those used on windlasses, winches, thrusters and starter motors are in fact only used for short durations. As there is a time factor in the heating of a cable, smaller cables can be used. The table shows battery cable ratings that are rated at 60% duty.

Battery Cable Ratings

Size AWG	Size B & S	Size mm^2	Rating (60% Duty)
8	8	8	90 amps
6	6	15	150 amps
4	3	26	200 amps
2	2	32	245 amps
1	0	50	320 amps
00	00	66	390 amps

Installing cable bundles

Where cables are bundled, the current carrying capacity must be reduced as the ability to dissipate heat is reduced. When more than 3 cables are bundled together in a large loom, the current capacity of the cable is reduced. The factor is typically around the nominal rating multiplied by 0.85 in commercial shipping standards and is based on 6 or above. The ABYC standards call for 30% reduction for 3 or more conductors, 4-6 a 40% reduction and 7-24 a 50% reduction. This is really only an issue in very large boats and where all conductors are under load. Small pleasure boats rarely have all under load together. Bundling reduces heat transfer from the conductors.

5. HOW TO INSTALL BOAT WIRING

How to install the cable runs

Cables should be neatly installed in as straight a run as practicable. Tight bends should be avoided to reduce unnecessary strain on conductors and insulation. The minimum cable bend radii of 4 x cable diameter is applicable to all cables but particular care should be taken with larger and more inflexible cables, and 6 x is a better target radius. The emphasis must be on accessibility, both for initial installation, maintenance and for the addition of circuits. Under no circumstances should you fiberglass in cables. All cables, in particular those entering transits, should be accessible to routine inspection. Connections should be minimized within any circuit between the power supply and the equipment. Connections and joins or splices in cables should be avoided. Any connection adds resistance to a circuit and introduces another potential failure point.

Protecting cables from mechanical damage

All cables should be installed to prevent any accidental damage to the insulation, or cutting of the conductors, or place undue strain on the cable. This may require protection within conduits or a protective batten or covering. In machinery or engine spaces, cables are often damaged during engine repairs and they should be covered where there is a risk.

Installing cables through decks and bulkheads

Cables passing through bulkheads or decks have to be protected from damage using a suitable non-corrosive gland, grommet or bushing. Cables transiting decks or watertight bulkheads should maintain the watertight integrity. Cable glands are designed to prevent cable damage and ensure a waterproof transit through a bulkhead or deck. A significant number of problems are experienced with the ingress of water through deck fittings and I have seen a variety of methods used. In addition running cables through GRP holes with some sealant invariably results in chafing and cable failure. Use circular multi-core cables if possible to ensure proper gland sealing is possible. The purpose designed Index (Thrudex) types are a good choice. The structural material of a deck has to be considered before selecting glands. A steel deck requires a different gland type than a foam sandwich boat.

Deck Cable Glands

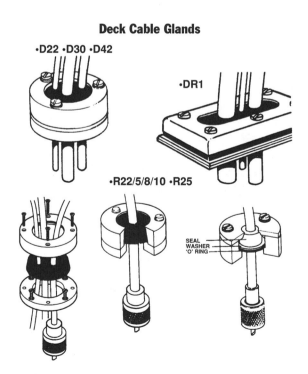

About cable clamp material

Cable saddles, straps, and cushion clamps should be of a non-corrosive material. Where used in engine compartments or machinery spaces, these should be metallic and coated to prevent chafe to the cable insulation. I prefer using standard electrical PVC-conduit saddles, which come in a variety of useful sizes available from any hardware outlet. The PVC cable tie or tie-wrap is universal in application, and should be used where looms must be kept together, or where any cable can be securely fastened to a suitable support. Do not use cable ties to suspend cables from isolated points as this invariably causes excessive stress and cable fatigue at the tie point. For internal cable ties, you only require the white ones. All external cable ties should be the black UV-resistant type.

About covering wires

If a number of cables are lying loose, consolidate them into some spiral wrap, and then fasten the loom using cable ties. The ABYC recommend using self-extinguishing coverings; an option is the use of split black tubing.

What about hot glue guns?

A hot glue gun is often very useful to fasten small or single cables above headliners when installing lighting circuits, or in corners behind trim and carpet finishes. It is useful where there is no risk of cables coming loose. However do not use on exposed cable runs or on heavy cables, without also fastening with mechanical brackets and saddles.

How far apart should cable clamps be?

Cables should be supported at maximum intervals of 8" (200mm). While the general ABYC recommendation is 18" (450mm) apart, I prefer the closer support distances to secure the cables more efficiently. I have seen far too much movement with the larger spacing distance, and sagging loops develop which are often easy to snag.

Installing the cable clamps

Cable saddles should fit neatly, without excessive force onto the cables, or cable looms, and not deform the insulation. Cables can be neatly loomed together and secured with PVC or stainless saddles to prevent cable loom sagging and movement during service. This is a common fault found on boats.

Installing wiring in lockers

In many vessels wiring must pass through lockers and small compartments used for gear storage. From experience I can say that most of the broken cable faults encountered are within such spaces. This happens when a variety of gear is thrown in and the impact severs or damages the wires. Always cover the wiring with split tubing or install within conduit to ensure mechanical protection. Also saddle the wiring loom well to limit movement and route the wiring to the top side edges of the compartments, to minimize the risks.

Installing heavy current cables

Heavy current consumers such as thrusters, windlasses, winches and toilet cables should be installed as far as practicable away from other cables. When large currents flow, interference may be induced into the other cables. Try and install them on the other side of the boat if possible.

Separating power, data and signal cables

Cables should be separated into signal, data or instrument cables, DC power supply cables, and where space allows heavy current carrying cables such as windlass or thrusters. This is to minimize induced interference between cables, in particular on long, straight parallel runs. All data and instrument cables should be routed as far as practicable away from power cables. Radio lead-in aerial cables should also be routed well way from power cables. AC and DC cables must be kept separated and never run together. A minimum distance of 12" (300mm) is recommended.

Installing instrument cables

Instrument and data cables are generally much smaller than power cables and do not have the same thick and robust insulation. They are easily damaged and care must be taken to avoid damage.

When data and power cables cross over

Cable crossovers are almost certain. Where instrument and data cables have to cross power cables, this should be done as close to an angle of 90° as practicable in order to prevent induced interference, with right angle crossovers. In addition spacing the cable apart as far as possible with a small air gap is good practice.

Cables Crossovers

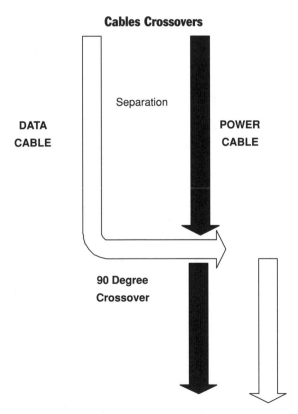

Separation

DATA
CABLE

POWER
CABLE

90 Degree
Crossover

Installing cables exposed to heat

Cables installed with machinery or engine spaces should be rated for the maximum heat of the space. In addition, where cables may be exposed to heat sources, such as exhaust manifolds or piping, they should be protected by conduits. It is also important to route cables away from heat sources or avoid the area altogether.

Installing cables in conduits

Where cables are installed within conduits or raceways, the conduits should be supported within 3" (75mm) of both entry and exit points. Conduit ends should be treated, or otherwise protected, to remove sharp edges and prevent chafe to cable insulation. Conduits are often installed during the construction phase, and this allows cables to be easily pulled in, replaced, or added. Conduits offer good mechanical protection to cables and in many cases, single-insulated cables are run in conduit back to the switchboard. As the cables are single insulated, they are exposed where they enter or exit the conduits, and should be supported by saddle or clamp to prevent excessive movement. Try to avoid installing large bunches of cables in flexible conduits as they tend to move around and chafe. PVC conduits should not be used in machinery spaces. Where cables exit conduits, the exit should be bushed to prevent chafing. During installations when pulling in cables, insulation is frequently damaged as insulation rubs against sharp edges.

Installing cables exposed to weather

All externally installed cables should be protected against the effects of ultraviolet (UV) light. Continued exposure to UV on external equipment cables will result in insulation degradation and failure. Small cracks in the insulation allow water to penetrate the conductor and subsequently corrode and degrade the copper. This is common on navigation lights, GPS aerial cables, radio aerial cables and other equipment. All exposed cables should be covered in fire retardant black UV resistant spiral wrapping or split loom to prevent rapid degradation of insulation. Cable ties should also be of the black UV type. Use tinned copper conductors on all external wiring to navigation lights, spotlights, and cockpit lights.

Grounding systems explained

It is very important to understand these definitions, as it is crucial to understanding their importance within respective circuits as well as to each other. Invariably confusion as to function is a key cause to system problems.

- **The DC negative**
  The DC negative is not a ground, but is a current carrying conductor that carries the same current that flows within the positive DC conductor.

- **The DC polarizing ground**
  In a single circuit wiring configuration, the battery is bonded to the mass of the engine by the negative conductor for the starting system. The engine is usually connected to an immersed item such as the steel hull or prop shaft. This is used to polarize the DC electrical system and doesn't actually carry current.

- **The negative to AC ground bonding link**
  Most recommendations (ABYC and ISO) call for a bonding link between the DC negative and the AC safety ground. The main stated justification is that if DC negative and AC ground are not connected and a short circuit condition develops between the AC hot conductor and a DC negative or bonding system, this would result in the AC protection not tripping. This may cause energizing of these circuits up to rated voltage, creating a risk to persons in contact. There are other stated risks of potentially fatal shock risks to swimmers. The second reason is to prevent potential difference arising that can cause corrosion. The controversy is considerable as this has resulted in many cases of increased corrosion rates, and also cases of the DC negative being alive with AC. The galvanic isolator is used in the AC ground conductor to DC ground point to eliminate the corrosion risk.

- The lightning ground

  The lightning ground is a point maintained at ground potential that is immersed in seawater. It only carries current in the event of a lightning strike and the primary purpose is to ground the strike energy. It is not a functional part of any other electrical system. While recommendations are to bond it to all other systems, my preference is not to interconnect it. This is to prevent high voltages being impressed on the DC electrical system. Several surveys revealed that serious damage was caused by this practice. and although a high voltage will be induced into wiring in a strike, the damage generally appears to be less. Radio grounding plates are not rated by manufacturers for use as lightning grounds and the use of the radio equipment ground is also not advised.

- The cathodic protection system ground

  The cathodic protection ground bonds together all items to be protected to main an equal potential or voltage level on them. This is then connected to the sacrificial zinc anode that is connected to protected underwater items. I prefer not to bond this to other systems to minimize effects that may increase or affect the protection system. In many cases anode corrosion rates are accelerated and when wastage is complete other parts of the boat start corroding if unchecked.

- The AC safety ground (or earth)

  The AC safety ground is a point at ground potential that is immersed in seawater. Under normal operating conditions, it carries no voltage or current. The primary purpose is that under fault conditions it will carry fault current to ground and hold all connected metal to ground potential, ensure operation of protective equipment, and protect people against electric shock from exposed metal parts.

- **The radio frequency (RF) ground**
  The radio frequency ground is an integral part of the radio aerial system and is sometimes termed the counterpoise. The ground only carries RF energy and is not a current carrying conductor. It is not connected to any other ground or negative. A dedicated ground plate or shoe is best practice. Some requirements allow a grounding plate used for radio also to be used as a lightning ground, but my advice is not to do this.

- **The instrument ground**
  The instrument ground, which most GPS and radar sets have, is nominally vessel ground. In many cases, a complete separate ground terminal link is installed behind the switchboard, and to which the cable screens and ground wires are connected. A separate large low resistance cable is then taken to the same ground point as other grounds. Do not simply interconnect the DC negative to the link as equipment may be subject to interference.

What is the DC grounding bus?

This term is generally applied to the use of a thick copper strip running the full length of the boat to join various grounds.

About grounding conductor screens

Screens should be grounded at one end only, or in accordance with specific manufacturers' recommendations.

About grounding electrical equipment

Some DC equipment have conductive metal cases. Where the equipment is installed in a wet location exposed to bilge or seawater, the casing should be grounded. The best method is to run a single insulated cable of equivalent size to conductors back to the ground link point. This recommendation is based on reducing electrolytic (stray current) corrosion and where a negative fault to case occurs.

About grounding electronics equipment

Electronic equipment grounds should be made back to the ground link point, or in accordance with specific manufacturers' recommendations. In many cases, a ground terminal block is installed close to instruments and a large ground conductor taken to the same point as the battery negative connection point. This is not to the battery negative but the actual termination point.

Grounds

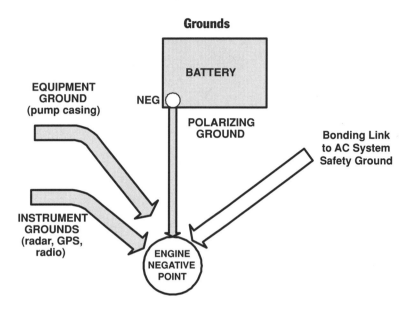

Installing the main negative connection

The main negative cable should be secured on the engine using a spring washer, to prevent the connection becoming loose through vibration. The main engine negative cable is prone to vibration from the engine. It frequently comes loose causing starting problems, intermittent equipment operation, interference, and in some instances alternator failures. In most cases, it is simply fastened to a convenient bolt. The mating surface must be cleaned to ensure a good electrical contact.

Avoiding magnetic loops

When the installed cable is routed close to a compass or other electromagnetic sensitive equipment, the term magnetic loop is often used. Always route well away from this type of equipment, which includes a magnetic compass, fluxgate compasses and electronics. A magnetic compass light usually has a twisted pair wire to the light, so make sure the connection is away from the compass and the supply cable is also installed well clear. It is not usually practical to install twisted pair or shielded cables near sensitive equipment.

6. MAKING WIRING CONNECTIONS

About crimp cable terminations

All conductors should be terminated where practicable using crimped connectors. Where cables are terminated within terminal blocks, they should be secured to prevent contact with adjacent terminals. The most practical and common method of cable connection is the tinned-copper, crimp terminal or connector. These are color coded according to the cable capacity that can be accommodated. Terminals are usually designed and manufactured according to NEMA standards, which cover wire pullout tension tests, and voltage drop tests. Where possible, select double crimp types, which should be used in high vibration applications.

Standard Cable Connectors Table

Color	AWG	Cable Sizes	Current Rating
Yellow	12-10	3.0 to 6.0 mm^2	30 amps
Blue	16-14	1.5 to 2.5 mm^2	15 amps
Red	22-18	0.5 to 1.5 mm^2	10 amps

Quick-disconnect (spade) connectors

These are commonly used, particularly on switchboards. Always select the correct quick-disconnect (spade) terminals for the intended cable size. Female connectors are easily dislodged, and have a tendency to slip off the back of circuit breaker male terminals, so ensure they are tight to push on. Ensure that the terminal actually goes on the CB terminal, and not in between the insulation sleeve and the connector. For hard duty, look at using heat-shrink fully insulated types. It is important not to apply too much strain on the cables.

Using ring terminals

Ring terminals are used on all equipment where screw, stud, bolt and nut are used. They should also be used on any equipment subject to vibration, or where accidental dislodging can be critical, particularly switchboards. Always make sure that the hole is a close fit to the stud, bolt or screw used on the connection, and to have good electrical contact and use spring washers.

In-line cable (butt) splices

Where cables require connection and a junction box is impractical, use insulated in-line butt splices. This is more reliable than soldered connections, where a bad joint can cause high resistance and subsequent heating and voltage drop. Use heat shrink insulation over the joint to ensure waterproof integrity is maintained. Some connectors when heated form a watertight seal by the fusing and melting of the insulation sleeve.

Using pin terminals

Pin terminals can make a neat cable termination into connector blocks. I have found these to be unreliable simply because vibration and movement work them loose. In most cases they do not precisely match the connector terminal and make a poor electrical contact.

Using snap plug (bullet) terminals

These are useful when used to connect cabin light fittings. I often use these on cable ends, female on the supply and male on the light fitting tails. This makes it easy to disconnect and remove fittings from head liners without having to install another crimp.

Can screw connectors be used?

Absolutely not. ABYC has a rule against their use. When screw connectors, or screw nuts are used, they screw the cable ends together, fatigue and often break the conductor strands, causing high resistance, heating and failure.

Making the crimp connection

When crimping, use a quality ratchet-type crimping tool. Do not use the cheap squeeze types, which do not adequately compress and capture the cable, leading to failure as the cable pulls out of the connector sleeve. A good joint requires two crimps. Always crimp both the joint and the plastic behind it. Ensure that no cable strands are hanging out. Poor crimping is a major cause of failure. A crimp joint can be improved by lightly soldering the wire end to the crimp connector. Avoid excessive heat. After crimping, give the connector a firm tug to ensure that the crimp is sound.

Ancor Crimp Tool
Courtesy ANCOR

Making wire terminations

Cable ends should have the insulation removed from the end, without nicking the cable strands. The bare cable strands should be simply twisted together, and inserted in the terminal block or connector of a similar size. Ensure there are no loose strands. Stray strands often cause short circuits with adjacent terminals. If you are terminating into an oversize terminal block, it may be more effective to twist and double over the cable end to ensure that the screw has something to bite on.

About solder terminations and connections

Do not solder the ends of wires prior to connection. In most cases, this is done to make a good low resistance connection and prevent cable corrosion. In my experience, soldered connections cause more problems than they prevent, The solder travels up the conductor causing stiffness, resulting in fatigue and early failure. In most cases, the soldering is poorly done, making a high resistance joint. A soldered cable end also prevents the connector screw from spreading the strands and making a good electrical contact, causing high resistance and heating. Use connectors of the correct size for the cable.

How to label wires

Always mark cable and wire ends to aid in correct reconnection and troubleshooting. The numbers should match those on the wiring diagram. A simple, slide-on number system can be used. The stick-on types should be avoided as they generally unravel and fall off as the adhesive fails. If wires are color-coded, use numbers, as they are easier and much quicker to identify. Commercial shipping use numbers, and the circuit positive should sequentially match the supply source such as the circuit breaker. The circuit negative should match the positive, and be placed in the same sequential order on the negative link. The numbering convention if unmarked is left to right.

Circuit Numbering

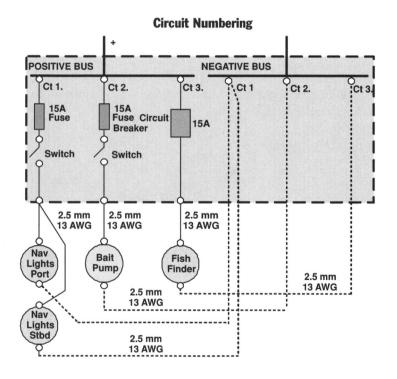

Installing plugs and sockets

Deck plugs and sockets are often used instead of deck glands and junction boxes at a mast base or as outlets for hand spotlights. Many are of inferior quality and prematurely fail. Don't use the cheap chrome plugs and sockets, they aren't waterproof. ABYC recommends snap closing covers although few are made this way. When using deck plugs ensure that the seal between deck and connector body is watertight. Leakage is very common on wet decks up forward where they are usually located. Make sure that the cable seal into the plug is watertight. It is of little use having a good seal around the deck, and plug to socket if the water seeps in through the cable entry and shorts out terminals internally as is often the case. Most connectors have O-rings to ensure a watertight seal. Check that the rings are in good condition, are not deformed or compressed, and seal properly in the recess. A very light smear of silicon grease assists in the sealing process. Ensure that the pins are dry before plugging in and that pins are not bent or show signs of corrosion or pitting. Do not fill around the pins with silicon grease, as this often creates a poor contact. Keep plugs and sockets clean and dry.

DriPlugs
CourtesyDri-plug

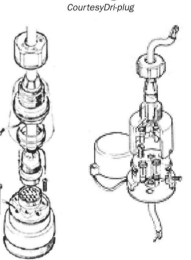

When to use cable junction boxes

Junction boxes are the most practical way to terminate a number of cables, especially where access is required to connect and disconnect circuits. To reduce the number of cables radiating back to the switchboard and minimize voltage drops, I use one junction box forward and one aft to power up lighting circuits. There are problems with using junction boxes. Terminal strips are installed internally and in many cases, the box is too small for the quantity or size of cables to be used. In these cases, cable tails to the block are made small which does not allow easy insertion. The box lid is forced on, applying pressure to the cables, This should be avoided as unnecessary stress is applied to terminations. Where there are many cables, use two junction boxes and split the circuits.

Junction Boxes
Courtesy Index Marine

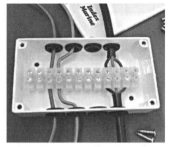

Installing wiring in junction boxes

Cables terminated within a junction box should enter from the bottom, and be looped to prevent water entering the box and connections. All cables should enter from the bottom. Junction box upper surfaces should have no openings that permit the entry of water. Cables looped in at the bottom will allow water to drip off, and prevent surface travel to the connections. Cables within junction boxes should be marked with numbers that correspond to circuit numbers used at the main switch panel. Cable ends should be numbered to aid in reconnection and troubleshooting. The numbers should match those on the wiring diagram.

Making connections in wet locations

Where connections are made within any area subject to water or moisture, such as bilges, the terminations should be made as far as practicable near or above the top of the maximum bilge level. Connections should be suitably protected against water ingress. Joints should be finished with self-amalgamating tapes, or some heat shrink tubing should be applied, preferably both. I have frequently seen connections for pumps and float switches that were permanently immersed in bilges fail. In automatic bilge circuits, the live connection also contributes to corrosion problems in some steel and alloy boats.

Testing the installation

Before powering up a circuit, all circuits should be tested to verify that the levels of insulation are satisfactory on the whole system, and on each circuit. Supply circuit breakers should be switched on so that the switchboard is included within the test. A multimeter set on the resistance range should be used between the positive and negative conductors. If readings are low, check that a load is not connected. The insulation resistance between conductors, or conductors and ground of all circuits or the complete installation should be greater than 100,000 ohms.

7. CIRCUIT PROTECTION AND ISOLATION

Why are control and protection required?

The heart of all electrical systems is the switchboard or panel, which allows control, switching and protection of circuits. Switches are used to isolate voltage from circuits as part of normal control. Protection is required to prevent overload currents arising in excess of the cable rating. They are also to protect the cables and equipment from excessive currents that arise during short circuit conditions. Circuit protection is not normally rated to the connected loads, although this is commonly done on loads that are considerably less than the cable rating, such as VHF radios or instrument systems. The two most common circuit protective devices are the fuse and the circuit breaker.

General System

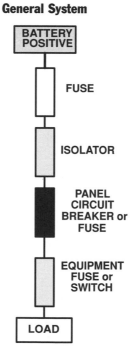

What is a short circuit?

A short circuit is where two points of different electrical potential are connected, that is positive to negative. There is what is often called a dead short circuit and an impedant short circuit. Dead short circuits are where the positive and negative are directly connected together without any resistance between them. This is actually rare and occurs when circuits are connected incorrectly, usually in a new installation. Impedant short circuits are the most common, where there is some impedance or resistance. Typical examples are where a breakdown across a terminal block occurs, or a winding.

What is an overload?

An overload condition is where the circuit current carrying capacity is exceeded by the connection of excessive load. Excessive load can come from too many devices or equipment such as pumps with higher than normal mechanical loads.

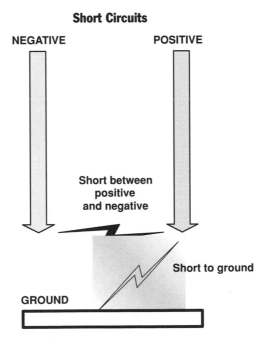

Short Circuits

NEGATIVE   POSITIVE

Short between
positive
and negative

Short to ground

GROUND

Mixing power and lighting circuits

Circuits should not have mixed consumers, such as power to out-lets or motors also connected to lighting equipment.

About circuit breakers

Circuit breakers are the most reliable and practical method of circuit protection. Physically they are manufactured in press button aircraft, toggle, or rocker switch type. They are either magnetic or thermal. They are used for circuit isolation and protection, combining both functions, which saves switchboard space, costs and installation time as well as improves reliability. Single pole circuit breakers are normally fitted to most vessels; however classification societies only allow these in grounded pole installations. This is because a fault arising on the circuit will provide a good ground loop and the large current flow will ensure proper breaker interruption. Double pole breakers are recommended for all circuits, as they will totally isolate equipment and circuits. This is a requirement of many classification or survey authorities.

Button Circuit Breaker   **Double Pole Circuit Breaker**   **Single Pole Circuit Breaker**

Courtesy Blue Sea Systems

What is a thermal circuit breaker?

These are the most common types and have a tripping mechanism with a thermal actuator and mechanical latch. This design allows discrimination between the initial in-rush and temporary surge currents and the longer period overloads. Typical applications are electric motors.

What is a magnetic circuit breaker?

These circuit breakers have a solenoid coil, and some have a hydraulic time delay. The solenoid coil has a non-magnetic delay tube inside. This contains a spring-biased and moving magnetic core. The armature is linked to the contacts and the coil mechanism. This functions as an electromagnet. When the contacts open, no current will flow through the circuit breaker, and the coil generates no electromagnetic energy.

Thermal Circuit Breaker
Courtesy Blue Sea Systems

What is a thermal magnetic circuit breaker?

It combines both types that have a solenoid connected in series with a bimetal thermal actuator. This gives a two-stage time and current tripping characteristic. In high over current short circuit situations, the solenoid trips quickly. The thermal function gives a slower trip function.

What does trip-free mean?

This means that a circuit breaker cannot be held closed or placed in the on position with an overload or overcurrent present.

What is interrupt capacity?

This is defined as the maximum fault current level that can be interrupted safely without the breaker failing.

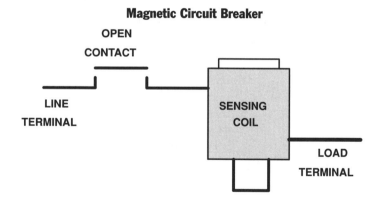

Magnetic Circuit Breaker

Selecting circuit breakers.

Circuit breakers must be selected for the cable size that they protect. The rating must not exceed the maximum rated current of the conductor. The cable sizes in the following table give recommended ratings for single cables installed in well-ventilated spaces. Ratings are given according to IEC Standard 157. Each trip time delay type can be illustrated in a trip time delay curve. These curves are used to show the relationship between the trip time in seconds and the percent of rated current.

What is a tripping characteristic?

Characteristics are normally given by the manufacturer of the breaker in a curve of current against time. The greater the current value over the nominal tripping value, the quicker the circuit breaker will trip. In cases of short circuit, tripping is rapid due to the high current values. Slower tripping characteristics are seen where a small overload exists and tripping occurs some seconds or even minutes after switch on. This happens as the current levels gradually increase.

Circuit breaker curve
Courtesy ETA

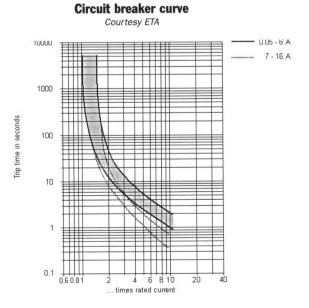

Circuit Breaker Selection Table

Circ Mils	AWG	Wire mm²	Current
3260	15	1.5 mm²	7.9 - 15.9 A
5184	13	2.5 mm²	15.9 - 22.0
8226	11	4.0 mm²	22.0 - 30.0
13087	9	6.0 mm²	30 0 - 39.0
20822	7	10.0 mm²	39.0 - 54.0
33088	5	16.0 mm²	54.0 - 72.0
52624	3	25.0 mm²	72.0 - 93.0
66358	2	35.0 mm²	93.0 -117.0
105625	0	50.0 mm²	117.0 -147.0

Circuit protection device approvals

All protection and isolation devices should have an assigned DC fault rating and be approved by a relevant national or international standard. Only install circuit breakers that are approved by UL, CSA or Lloyd's. Approvals for small boat breakers categorize them as supplementary protectors. I normally use ETA, Ancor and Carling circuit breakers. They must be approved for DC operation and be marked with the rating.

CE Approval Logo

(C€ *logo denotes compliance with European Standards. Technical Construction files for C€ marked products on file at Blue Sea Systems*)

UL Approval Logo

LISTED

How does a fuse work?

Fuses are still commonly used, and are either housed in simple fuse holders or combination fuse switches. The principal advantage is the lower capital cost. A fuse is essentially a strip of a low melting point alloy within a housing that is placed in series with a circuit. When the current increases under load and exceeds the rating during an overcurrent event such as an overload or short circuit, the fuse element melts, or also blows or ruptures. The time it takes for a fuse to blow is dependent on the current level as a ratio of the fuse rating. If a very large fault current rating is flowing, at 300% it may be less than a second, at 200% it may take 5 seconds, at 150% it may take 2 minutes.

Blade Fuses

Blade Fuse panel
Courtesy Blue Sea Systems

What types of fuses are used?

The most common type of fuse is the simple glass fuse. AGC and GMA Fuses have a wire type fuse element, AGU fuses have a thin flat element, the MDL Slow-Blow fuses have a thin spiral type element, and the ATO/ATC and ATM fuses are the automotive blade fuses contained within color coded plastic fuse holders. The MAXI fuse is a larger version of the ATM types.

AGC Fuse   **MDL Fuse**

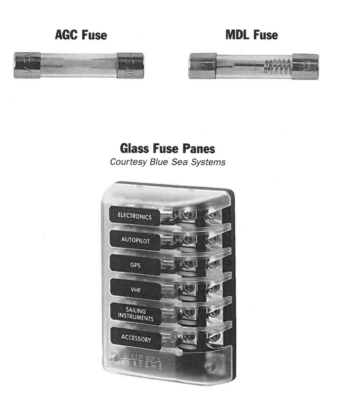

Glass Fuse Panes
Courtesy Blue Sea Systems

What are slow blow fuses?

Slow blow fuses are those that have an element designed to take longer to blow. This is to prevent nuisance blowing when very short duration or transient overloads occur with the equipment starting up or under loads. Anchor windlasses are typical. For example at 150% it takes 100 seconds, at 200% it takes 10 seconds, and 300% it takes 1 second.

What are the large fuses used for?

These are called Mega, Class T and ANL fuses. These fuses are used for larger current ratings in the range of 100 to 500 amps. Typical circuits are thrusters, inverters and windlasses. The fuses are bolted onto studs in a fuse block carrier. The fuses also have an indicator that shows when a fuse is ruptured. These fuses are required to have insulating and protective covers to meet USCG and ABYC recommendations.

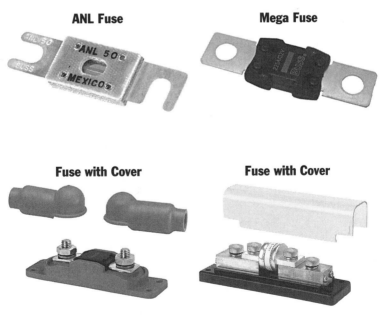

ANL Fuse   **Mega Fuse**

Fuse with Cover   **Fuse with Cover**

Courtesy Blue Sea Systems

What are the problems with fuses?

The typical glass fuse is not always accurate and can rupture as much as 10-50% above or below nominal current rating. Fuse elements can fatigue in service with the properties altering and subsequently the rated value may change, causing nuisance failure. Vibration also commonly causes failure. There is often added contact resistance in the fuse holder between each contact and the fuse ends which commonly causes voltage drops, intermittent supply and heating. When a fuse fails, always assume there is a fault.

Protection of switchboard circuits

The power supply to the switchboard must have short circuit protection. This is usually a slow blow fuse in the 100-200 amp rating. Circuit isolation must be installed as close as practicable to the battery in both the positive and negative conductors. This is a requirement of ABYC, although it is not always possible. The isolator should also be accessible. It may be incorporated within a single trip free circuit breaker. The isolator must be rated for the maximum current of the starting circuit. Fuses can be used; however, it is often preferable to combine isolation and protection within one easily re-settable device. They should also be mounted as high as practicable above possible bilge and flooding levels.

What is protection discrimination?

This term may also be known as protection coordination. The principle of discrimination is used in DC and AC circuits. A circuit will normally have two or more over-current protective devices, such as the main and auxiliary circuit breakers installed between the battery and the load. The devices must operate selectively so that the protective device closest to the fault operates first. If it doesn't, the second device will operate, protecting the circuit against over-current damage and possibly fire.

Protection Discrimination

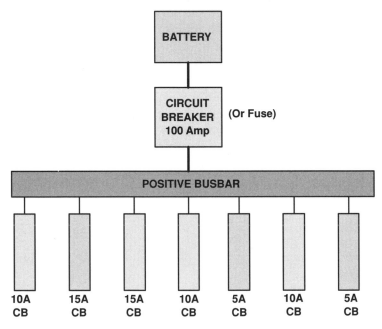

How do you install protection?

Use circuit breakers with different current ratings. This effectively means that at a point on the time delay curve the first breaker will trip. If it does not and the current value increases, the next will. A point is reached called the limit of discrimination. At this point the curves intersect and both breakers will trip simultaneously. Circuit breakers with different time delay curves can be used. This simply means using breakers with differing time delay curves to achieve the same result. Finally you can use circuit breakers with different time delay curves, current ratings and different breaker types. This enables using all of the above to ensure discrimination.

Installing switchboard panel feeder protection

The ABYC recommends installing the circuit protection device such as circuit breaker as close as possible. As feeder cables are generally single insulated, this would mean 7 inches or less. On most vessels having an accessible circuit breaker would be impossible, so place a split conduit sheath over the cable and locate it as close as practicable and accessible as you can.

What about motor internal protection?

There is an ABYC requirement for motors to have internal protection against an overcurrent condition. Many motors do not have this imbedded temperature switch. This requirement is most common in the US. Ensure that the motor supply conductor is properly protected for the over current.

Protecting navigation lights

A circuit breaker supplying all navigation lights has the risk of a single fault tripping the breaker and all lights being unavailable until the fault is cleared. This may not be possible in adverse weather conditions. Where possible, separate circuit breakers or fuses should be used. Alternatively where a single breaker is used, each circuit should have a replaceable fuse and switch installed. This may be a multi-circuit fuse block or the rear of the switchboard, carrying fuses or circuit breakers for all circuits.

Selecting good isolators is essential

The isolator is usually a switch. It is recommended that you use quality single pole isolators. Do not use the 2 position selector switches, or cheap imported isolators that look like the name brands. I have had many failures with these copies. Installing quality will pay dividends. If you lose the main isolator you lose everything so reliability is essential.

Courtesy Blue Sea Systems

What is a remote isolator?

This is a method for isolating the power when circuit breakers and isolators are not accessible. The isolator is installed in line close to the battery. A pair of control cables are connected and run to a switch located in a suitable location. An ETA isolator is shown.

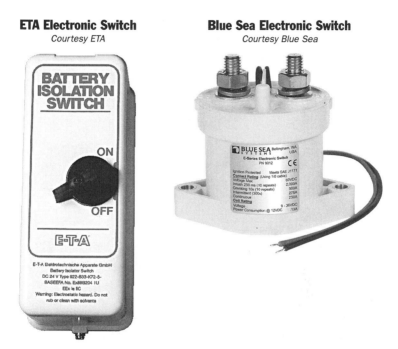

ETA Electronic Switch
Courtesy ETA

Blue Sea Electronic Switch
Courtesy Blue Sea

About busbars and terminal blocks

All fuses, distribution busbars, and terminals must be covered. The insulative covers should be fitted over all positive and negative busbars, distribution busbars and fuse holders. This is a requirement of ABYC and protects against accidental contact and water.

Link Bar with Cover

Negative Link Bar

Single Pole Terminal Insulated

Courtesy Blue Sea Systems

8. ABOUT SWITCHBOARDS

What should switchboards be made of?

The switchboard or panel shall be constructed of non-hygro-scopic (non-water absorbing) and fireproof material. The panel should be rated to a minimum of IP44. Switchboard panels are normally made from aluminum, or plastic based materials. Ideally panels should be non conductive; however many are made of etched aluminum. They should be rated to meet either an IP or NEMA standard against the ingress of water.

Breaker Panel with Digital Meter
Courtesy Blue Sea Systems

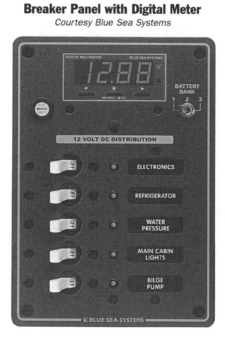

Fireproofing switchboard interiors

The switchboard interior walls should be fireproof or incapable of supporting combustion. Many survey authorities specify that the internal part of the switchboard must be lined with a fire resistant lining. Line all interior walls with appropriate sheeting. This will help in containing any fire that may arise in severe fault and fire conditions.

Where switchboards should be located

The switchboard should be located in a position to minimize exposure to spray or water. Where occasional spray is possible, some protection is recommended, which may be a clear PVC cover or similar measure.

About mixing DC and AC wiring systems

DC systems should not be located or installed adjacent to AC systems. Where DC and AC circuits share the same switchboard, they should be physically segregated and partitioned to prevent accidental contact with the AC section. The AC section must be clearly marked with Danger labels. The DC switchboard should not be integrated with the AC system. Where possible the AC panel should be located in a different location. This eliminates the chances of accidental contact with live circuits, or confusion between wiring systems. Where systems are integrated, physical separation should be used to prevent contact. The barriers should be well marked warning of the danger.

Switchboard load assignment

In many switchboards there are often 2 to 3 vertical rows of circuit breakers or fuses. When planning the board, consider installing one vertical bank to electronics equipment and the other to power equipment. This can reduce some interference. Each busbar side can be given a separate power supply as described in the following paragraph.

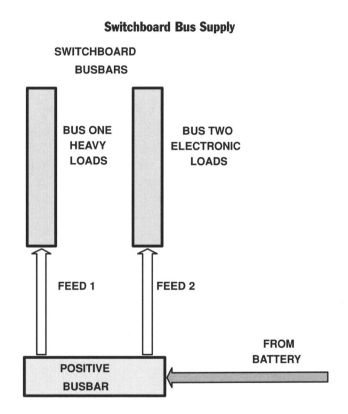

Switchboard Bus Supply

About switchboard busbars

The switchboard busbars normally consist of a copper bar that connects all the input terminals of circuit breakers or fuses. These bars are installed on vertical rows of circuit breakers. Normally busbars are series connected with one main positive input cable connected to a terminal at one end of the busbar. It is recommended that a separate power supply cable be installed to each busbar. This gives some redundancy if the connection should fail. When the main battery supply comes to the switchboard, it is good practice to terminate it on a large terminal block. Each busbar supply can then be supplied through two smaller supply cables.

About negative busbars

Each circuit negative should be terminated within a link or busbar. The number of the termination should be the same as the circuit breaker number. The battery negative is also terminated in the busbar.

What is the purpose of a voltmeter

A voltmeter should be installed to monitor the voltage level of the service and starting battery. A voltmeter will also tell you if the battery is charging at the correct voltage level. As a battery has a range of approximately one volt from full charge to discharge condition, accuracy is crucial. Analog voltmeters are the most common. A switch may be installed to enable monitoring of the service and start batteries from the same meter. Digital voltmeters are relatively common and are far more accurate. They are susceptible to voltage spikes and damage and many have maximum supply voltage ranges of 15 volts. There are a number of types, including Liquid Crystal Displays (LCD) and Light Emitting Diodes (LED). Where one voltmeter is used to monitor two or more batteries, switching between batteries to voltmeter is through a double pole, center off toggle switch or a multiple battery rotary switch. The sense cable should go directly back to the battery, although on service battery connections most connect directly to the switchboard busbar. Switch off the meter after checking.

Analog Voltmeter
Courtesy Blue Sea Systems

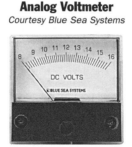

Protecting the voltmeter.

As a voltmeter is connected across the supply, that is positive and negative, protection against short circuit is required. This is done by installing fuses on either the positive or both input wires.

Digital Voltmeter
Courtesy Blue Sea Systems

Voltmeter Connection

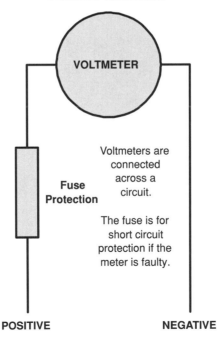

Fuse Protection

Voltmeters are connected across a circuit.

The fuse is for short circuit protection if the meter is faulty.

POSITIVE   NEGATIVE

What is the purpose of an ammeter?

An ammeter should be installed to monitor the discharge current rate from the service battery. The installation of an ammeter in the primary charging circuit to monitor the charging current is recommended. Ammeters are installed on the switchboard input positive to monitor service battery discharge current levels. Ammeters should be selected on the calculated operating range. An ammeter on the charging system can indicate that current is flowing. Cheaper ammeters are of the series type with the cable under measurement passing through the meter. There is often very long cable runs with resultant voltage drops in the charging circuit, and if the meter malfunctions damage can occur. Preference should be given to a shunt ammeter. The digital ammeter often uses a different sensing system. Instead of a shunt, the digital ammeter has what is called a Hall Effect sensor on the cable under measurement. The Hall Effect transducer generates a voltage proportional to the intensity of the magnetic field it is exposed to. For boat applications a 0-10 volt transducer output corresponds to a 0-200 amp current flow. Sensitivity is increased, and range reduced by increasing the number of coils through the transducer core.

Analog Ammeter
Courtesy Blue Sea Systems

What is a shunt ammeter?

A shunt simply allows the main current to flow through a resistance. This produces a voltage drop and this millivolt value is directly in proportion to the current flowing. The rating or ratio of a shunt will be given in the mV to amp ratio, i.e. 50mV/50A, meaning that 50mV will equal 50 amps. The meter reads the millivolt value and is displayed on an amp scale. The advantage is that only two low current cables are required to connect the ammeter to the shunt. Do not run the main charging cables to where the meter is and connect it, as this can insert excessive voltage drop into the charging circuit. Install a shunt in the line wherever practical and run sense wires back to the panel mounted meter. Always connect the meter leads to the special connection screws.

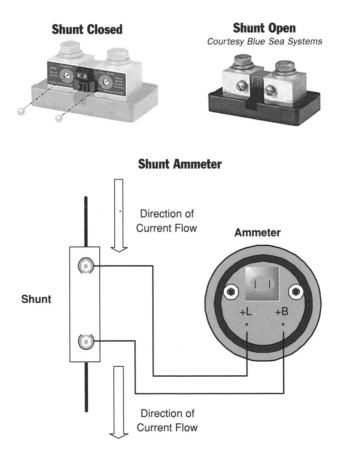

Shunt Closed

Shunt Open
Courtesy Blue Sea Systems

Shunt Ammeter

Direction of
Current Flow

Ammeter

Shunt

+L   +B

Direction of
Current Flow

About smart meters and monitors

Unlike starting batteries, house battery charge levels cycle up and down, and power level information is critical in determining charging periods. Typical of integrated monitors is the E-meter (Link 10). These "intelligent" devices monitor current consumption and charging current. They also have a range of monitoring functions that includes voltage, high and low voltage alarms, amp-hours used and amp-hours remaining. This allows the battery net charge deficit to be displayed. The system also maintains accuracy by taking into account charging efficiency. The charging efficiency factor (CEF) is nominally set at 87%, with the factor being automatically adjusted after each recharge cycle. A falling CEF is indicative of battery degradation. In addition the E-meter also contains a 'n' algorithm for calculation of Peukerts coefficient. A meter shunt (500A/500mV) is installed in the negative load line. It is connected by twisted pair wires to prevent noise from induced voltages being picked up and carried into the meter, corrupting data.

Smart Meter
Courtesy Mastervolt

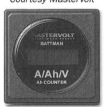

E-Meter/Link 10 Wiring

Identification of protection devices

All circuit protection and isolation or control devices should be properly labeled to allow easy identification. This should also include the circuit number if practicable.

Indicating circuit status

All circuit isolation and protection devices should have visual status indication, for OFF and ON. Circuit breaker status indicators consist normally of LED lights, filament lights, or backlit nameplates. Generally Green indicates off and Red is on. An LED requires a resistor in series. Red filament lamps are also commonly used. The one disadvantage of these is that they consume power, typically around 40 mA. If there are twenty circuits on, this adds up to a reasonable load on the system, and a needless current drain. If you have a very large switchboard, allow for the current drain. In many cases people assume they have a current leakage problem when in fact it is the switchboard indicators causing the drain. In many cases the location of the switch, either up or down, or left or right indicates whether the circuit is on. At night light indicators are easier to observe.

Switchboard cable installations

All cables, and cable looms to switchboard panels should allow the opening of the panel without placing strain on connections or cables. In many installations, cable looms are too short to allow easy opening of panels for inspection. It is common to have the connectors pulled off the rear of circuit breakers due to the strain on conductors or wiring looms. Looms should be neatly tied and in general, this may be in 2 or 3 smaller and separate looms. They should have sufficient length to allow complete opening of the switchboard, and the circuit cables should be secured to prevent undue stress on the connectors.

About switchboard troubleshooting

There are a number of faults that routinely occur on switchboards, and their protective devices. The following faults and probable causes should be checked first. It is assumed that power is on at the switchboard.

The circuit breaker is tripping when switched on

The ammeter shows an off-the-meter, full-scale deflection which indicates a high fault current:

- **Load short circuit.** Check out the appropriate connected load and disconnect the faulty item before resetting.

- **Connection short circuit.** If after disconnecting the load the fault still exists, check out any cable connections for short circuit, or in some cases cable insulation damage.

The circuit breaker trips several seconds after switch on

The ammeter shows a gradual increase in current to a high value before tripping off, and is typically an overload condition.

- **Motor is seized.** This fault may happen if the electric motor has seized, or more probably the bearings have seized.

- **Load seized or stalling.** This fault is usually due to a seized pump, mostly bearings.

- **Insulation leakage.** This fault is usually due to a gradual breakdown in insulation, such as wet bilge area pump connection.

There is no power after circuit breaker switch on

If power is absent at the equipment connection terminals, check the following:

- **Circuit connection.** Check that the circuit connection has not come off the back of the circuit breaker. Also check the cable connection to the crimp connection terminal.

- **Circuit breaker connection.** On many switchboards, the busbar is soldered to one side of all distribution circuit breakers. Check that the solder joint has not come away. In some cases breakers have a busbar that is held under breaker screw terminals. Check that the screws and connection are tight.

- **Circuit breaker.** Operate the breaker several times. In some cases the mechanism does not make proper electrical contact and several operations usually solve the problem by wiping the contacts.

- **Circuit negative.** If all tests verify that the positive supply is present, check that the circuit negative wire is secure in the negative link.

Circuit power on but no indication light

The LED may have failed, and in some cases the resistor. Also check the soldered connection to the circuit breaker terminal.

Installing auxiliary circuits

The power supply to auxiliary equipment connected directly to the battery must have short circuit protection and circuit isolation installed as close as practicable to the battery in both the positive and negative conductors. This may be incorporated within a single trip free circuit breaker. These auxiliary supplies generally include high current equipment such as thrusters, electric windlasses, winches, toilets, etc. connected directly to a battery. A circuit breaker rated for the cable should be installed as close as possible to the battery, and be accessible. They should also be mounted as high as practicable above potential bilge and flooding levels. Avoid running smaller circuits and equipment off batteries as control and monitoring are much harder to achieve.

Auxiliary Circuits

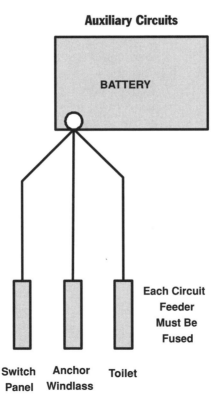

9. BILGE PUMP SYSTEMS

About bilge pump wiring

Where bilge pump control circuits incorporate automatic operation (e.g. float switch), caution should be given to the risks of pollution by uncontrolled or unmonitored discharge of oily bilge water. Where pumps are running in this mode, you should be aware that uncontrolled discharges of oily bilge water into the water might render you liable for stringent penalties and fines. A bilge pump automatic circuit is controlled by a 2-position and center off switch. In position 1 the power is applied directly to the bilge pump. In position 2 the power is supplied to a float or other switch. When the water level is low and the switch is not activated, power goes to one side of the switch. When water raises and the switch activates and closes, power is then taken directly to the pump. The correctly rated switch must be installed for the installed pump. It is common to see a 5-amp switch installed on an 8-amp pump. When operating at full load a large volt drop develops and the pump capacity is effectively and dangerously de-rated. Bilge pump wiring is generally in the bilge area and the connections must be waterproof and located above maximum water levels.

Bilge Pump Circuit

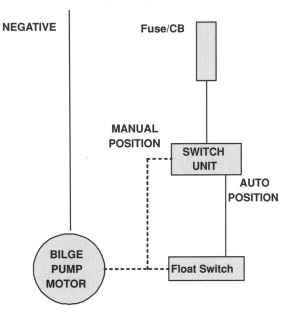

NEGATIVE

Fuse/CB

MANUAL
POSITION

SWITCH
UNIT

AUTO
POSITION

BILGE
PUMP
MOTOR

Float Switch

Bilge Control Switch
Courtesy Blue Sea Systems

Installing bilge alarms

The main bilge should have a separate visual and audible alarm to indicate levels above the normal operating range of automatic bilge pump systems. Most boats have automatic bilge pumping systems that use a float switch. If the pump is running and cannot keep up with the water, an additional alarm will indicate the high bilge level. If the float does not operate and the bilge starts to fill, a separate alarm will also indicate the condition.

Bilge Alarm Circuit

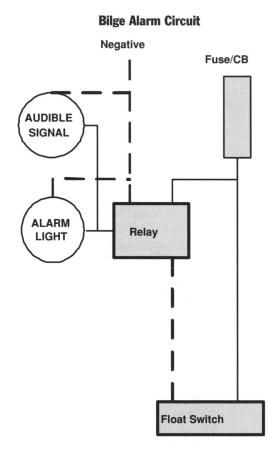

10. INSTALLING MAST AND EXTERNAL WIRING

Installing mast wiring

Mast wiring is a common source of failure. Many problems can be avoided if cables are installed properly. Since masts are generally wired by mast manufacturers and riggers, boat owners rarely take the opportunity to supervise or specify requirements. The major problem is the use of single insulated untinned cables, generally of an under-rated conductor size. Small conductor sizes cause many voltage drop problems with unacceptable low light outputs as a result. I recommend a 15-amp rated cable for each circuit.

Mast wiring negative conductors

Masthead tricolor navigation lights are usually integrated with an anchor light fitting. These use a 3-wire common negative arrangement. The same arrangement is often used for combination masthead and foredeck spotlights. The purpose is to save on one conductor, and also only one light is on at any one time. If two lights are on together, the negative will be under-rated for the current. Never use the mast as a negative path return as I have found on some vessels. Install a negative wire to each light fitting.

Protecting mast cables against UV

All exposed cables should be covered in black, UV-resistant spiral wrapping to prevent rapid degradation of insulation. Small cracks in the insulation allow water to penetrate the conductor and subsequently anneal the copper. This should be applied where a cable exits the mast to a light fitting.

Supporting internal mast cables

Cabling must be properly secured within the mast. The weight of a cable hanging down inside a mast causes fatigue through stretching. If the cables are not enclosed in a conduit, which is still relatively common, the internal halyards can whip against them, often severing conductors in multi-core instrument cables or severely damaging light cable insulation. There are a number of methods that can be used to secure mast cables; a combination of all three is best. Where cable enters the mast base and exits at the masthead it should pass through a cable gland. The ideal units for this are the Thrudex DR1 rectangular units. Once cables have been placed through the neoprene, the gland is tightened and compression around the cables takes the strain. An additional advantage is that cables are properly protected from chafe and damage against the mast entrance hole, which often have sharp edges. A small nylon blind cord messenger line can be installed with the cables, and supported at the masthead. The messenger should be tied or taped to the cable loom and then fastened to take the load off the cable ends. An obvious advantage is that the messenger serves as a pull through for cable additions or replacements. The very big disadvantage is that once the loom is taped to the line over the entire length, it is impossible to remove and replace single cables. Where possible, cable ties should be used to fasten and support cables. The ideal place to do so is where cables come out of the mast to connect lights, radar, etc. which usually gives 3-4 fastening points. There is generally sufficient space to insert a tie around the cables. A second hole large enough for a tie is required next to the main cable entry to enable tie to be supported.

Mast base junction boxes

The most common failure area is the junction box. If mounted inside the vessel, a good water resistant box should be installed. If mounted externally, and this should always be a last resort, a waterproof box is required. Always leave a loop when inserting cables into the box. If water does travel down the loom, this will drip off the bottom of the loop and will not enter and corrode the junction box terminals or connections.

Mast wiring deck cable transits

Cable glands are designed to prevent cable damage and ensure a waterproof transit through a bulkhead or deck. A significant number of problems are experienced with the ingress of water through deck fittings. I have seen some amazing systems utilizing pipes, hose, etc. If figure 8 type cable is used, or small, single insulated cables installed, it is very difficult to adequately seal them in cable glands. To overcome this problem, use circular multi-core cables if possible or consolidate the cables to make a cable loom that can be put through a deck gland. Some types are designed such as those from Index (Thrudex). You need to consider the structural material of a deck before selecting a gland. A steel deck requires a different gland type to a fiberglass, foam-sandwich boat.

Using deck plugs for mast wiring

Deck plugs are often used instead of deck glands and junction boxes at a mast base. Many plugs and sockets are of inferior quality and prematurely fail, generally when you need them most. Don't use the cheap and nasty chrome plugs and sockets, they aren't waterproof. When using deck plug, make sure that the seal between deck and connector body is watertight. Leakage is very common on wet decks up forward where the plugs are usually located. Make sure that the cable seal into the plug is watertight. It is of little use having a good seal around the deck, and plug to socket if the water seeps in through the cable entry and shorts out terminals internally as is often the case. Most connectors have O-rings to ensure a watertight seal. Check that the rings are in good connection, are not deformed or compressed, and seal properly in the recess. A very light smear of silicon grease assists in the sealing process. Ensure that the pins are dry before inserting the plug, and check that the pins are not bent or showing signs of corrosion or pitting. Do not fill around the pins with silicon grease, as this often creates a poor contact. Keep plugs and sockets clean and dry.

How to test and maintain mast cables

The mast subjects cables to all of the worst damaging factors, such as vibration, exposure to salt water, stretching and mechanical damage. There are basic maintenance tasks that will reduce mast wiring problems. Regularly examine cables where they exit the mast for signs of chafe. If the cable loom has not been protected with a UV-resistant sleeve, carefully examine the insulation for cracks. Regularly examine masthead cable exits for chafe. Ensure that coaxial, wind instrument and power cables have a reasonable loom to allow for shortening and repair.

Testing to Mast Circuits

TEST

Tricolor Positive, Anchor Positive, Common
Negative, Spreader Lt Positive
Spotlight Positive, Running Lt Positive

All Negatives

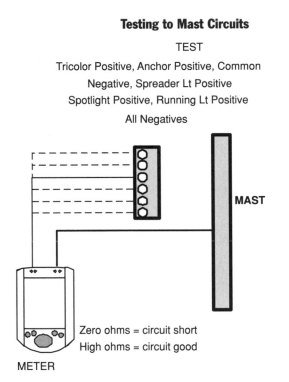

MAST

Zero ohms = circuit short
High ohms = circuit good

METER

Troubleshooting tri-color/anchor lights

If a light does not illuminate, lamp failure will be the usual cause as masthead vibration is a major factor. If the lamp is replaced and it still does not come on, open the mast connection box and locate the appropriate terminals. Use a multimeter set on the DC-volt range, and check that voltage is present at the terminals with power on. Many failures are due to poor contacts within terminal blocks, or corrosion of the terminal and cable. To perform a cable continuity test turn the power off, and with a multimeter set on the resistance x1 range, test between the positive and negative terminals. The reading should be approximately 2-5 ohms with known good lamp installed. If the reading is over-range, the light fitting or connection has failed or the cable has been damaged. The mast cable entry and exit points should be examined first. Internal breaks only occur in masts without wiring conduits. Many tricolor-anchor lights have a plug and socket arrangement, and these are an occasional source of failures.

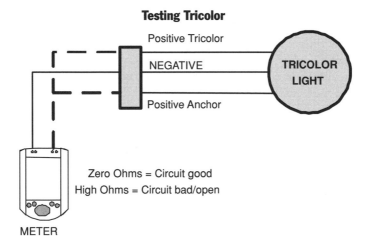

Testing Tricolor

Positive Tricolor

NEGATIVE

TRICOLOR LIGHT

Positive Anchor

Zero Ohms = Circuit good
High Ohms = Circuit bad/open

METER

How to troubleshoot mast spreader lights

The above tests are also valid for spreader lights. On many vessels, spreader lights are a sealed beam unit in a stainless steel housing. It is very common to have short circuits to the mast, as cables chafe through on the sharp edges. This problem is common for circuit leakages and increased corrosion rates on steel vessels:

- **Mast short circuits.** With a multimeter set on the resistance ohms x1k range check between mast and both the positive and negative wires. The reading should be over-range. If you have any reading you have either a short or a leakage from cable insulation breakdown.

- **Check supply.** Open the mast connection box and locate the appropriate terminals. Using a multimeter on the DC-volt range, check that voltage is present at the terminals with power on.

About running fly bridge cables

When running fly bridge electrical and electronics cable, these are usually installed within the stainless tubing structure. There are often cable faults when the wires are cut or chafe on sharp edges where they enter or exit the tubing. The sharp edges where tubing is joined are another unseen hazard, so always use quality double insulated tinned cables.

About installing external wiring

In many cases navigation lights and instrument aerials are installed and mounted on stainless steel stanchions. The wiring is run through the deck and the cable internally within the stainless tubing. The two main failure points are where the cables run through the deck and where cables enter and exit the stainless tubing. Many deck cable glands are located in exposed areas, causing mechanical stress and damage. Always place them as far behind tubing base fittings as practicable. In many cases this is difficult as the deck to hull joint is located in the same location and location is limited by below deck cable access. Where cables enter and exit, they should be protected by a sleeve or grommet. A common cause of failure is cable and insulation damage from sharp edges on the holes drilled into the stainless steel tubing. As previously mentioned, apply UV protective coverings to exposed cables.

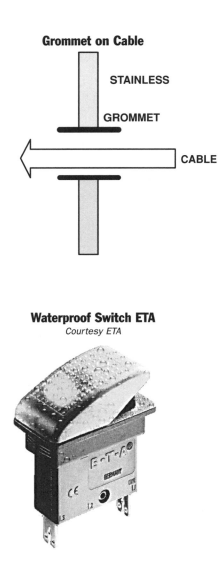

Grommet on Cable

STAINLESS

GROMMET

CABLE

Waterproof Switch ETA
Courtesy ETA

Acknowledgements

Thanks and appreciation go to the following companies for their assistance and readers are encouraged to contact them for equipment advice and supply. Quality equipment is part of reliability!

ANCOR	www.ancorproducts.com
Blue Seas	www.bluesea.com
Bulgin	www.bulgin.com
Carling	www.carling.com
Dri-plug	www.dri-plug.com
Index Marine	www.indexmarine.co.uk
NewMar	www.newmar.com
Marlec	www.marlec.co.uk
Mastervolt	www.mastervolt.nl
Xantrex	www.xantrex.com

The Marine Electrical School

This book contains the material for Module 101 of the Certificate in Basic Marine Electrical and Electronics. Log on to www.marineelectrics.org for course details.

Index

Other books by John C. Payne:

MOTORBOAT ELECTRICAL & ELECTRONICS MANUAL
by John C. Payne

Following the international success of *Marine Electrical and Electronics Bible*, Payne turns his talents from sailing boats to powerboats. This complete guide, which covers inboard engine boats of all ages, types, and sizes, is a must for all builders, owners, and operators. Payne has put together a concise, useful, and thoroughly practical guide, explaining in detail how to select, install, maintain, and troubleshoot all electrical and electronic systems on a boat.

Contents include: diesel engines, instrumentation and control, bow thrusters, stabilizers, A/C and refrigeration, water and sewage systems, batteries and charging, wiring systems, corrosion, AC power systems, generators, fishfinders and sonar, computers, charting and GPS, radar, autopilots, GMDSS, radio frequencies, and more.

"... tells the reader how to maintain or upgrade just about every type of inboard engine vessel." *Soundings*

MARINE ELECTRICAL & ELECTRONICS BIBLE
by John C. Payne

"Everything a sailor could possibly want to know about marine electronics is here...as a reference book on the subject it is outstanding."
Classic Boat
"A bible this really is...the clarity and attention to detail make this an ideal reference book that every professional and serious amateur fitter should have to hand." *Cruising*
"...this is, perhaps, the most easy-to-follow electrical reference to date."
Cruising World
"All in all, this book makes an essential reference manual for both the uninitiated and the expert." *Yachting Monthly*
"...a concise, useful, and thoroughly practical guide.... It's a 'must-have-on-board' book." *Sailing Inland & Offshore*

SHERIDAN HOUSE
America's Favorite Sailing Books
www.sheridanhouse.com